Birds, Beasts & Reptiles

The Natural History Etchings of Mark Catesby

A Book of Postcards

The Huntington Library

Pomegranate
SAN FRANCISCO

Pomegranate
Box 6099
Rohnert Park, CA 94927

Pomegranate Europe Ltd.
Fullbridge House, Fullbridge
Maldon, Essex CM9 4LE
England

ISBN 0-7649-0368-3
Pomegranate Catalog No. A911

Pomegranate publishes books of
postcards on a wide range of subjects.
Please write to the publisher for more information.

Designed by Shannon Lemme
Printed in Korea
06 05 04 03 02 01 00 99 98   10 9 8 7 6 5 4 3 2

Mark Catesby (English, 1682–1749) embarked on his first major venture as a naturalist in 1712, when he accompanied his sister, Elizabeth, to Williamsburg, Virginia, where she was reunited with her husband. During his stay he explored the colony as far as the Blue Ridge and also visited Jamaica. Plants sent back to England established his reputation among the natural history circle of London.

In 1722, sponsored by a group of eminent naturalists and backed by the Royal Society, Catesby embarked on the journey that would result in *The Natural History of Carolina, Florida and the Bahama Islands* (1731–1743). He spent the next four years collecting and sketching in the Carolina low country and Piedmont, as well as the region that would soon become the colony of Georgia. He then explored the Bahama Islands.

Catesby's career as a field naturalist ended on his return to London in 1726; he spent most of the rest of his life producing *The Natural History*. The work, published by subscription, was to appear in ten sections of twenty plates each. An Appendix of twenty additional plates did not appear until 1747. Catesby learned to etch his own plates and produced almost all of the 220 illustrations in the work. He also wrote the accompanying text and oversaw the coloring of the plates.

Catesby's work expressed a newly emerging concern for the way in which plants and animals from the same habitats interrelate—an environmental understanding that moved a small group of innovative naturalists to consider the cosmos as an intricate matrix of interdependent parts. As such, Catesby's publication worked against the predominant urge of eighteenth-century scientific thinkers to classify flora and fauna as individuated forms in abstract taxonomic systems—an urge reflected, for example, in the work of Catesby's celebrated contemporary, Carl Linnaeus. Catesby's *Natural History* can be understood as a cornerstone in the development of the modern science of ecology, and his work lies at the beginning of an extended discourse which, over the course of two and a half centuries, has come to define the North American continent as a complex set of environmental systems.

This book of postcards presents thirty of Catesby's stunning images, all from the Huntington's copy of the first edition of *The Natural History*. The common names and Latin identifications from the original plates, along with the Linnaean index, have been taken from George Edward's 1771 edition of Catesby's work.

T 83

Birds, Beasts & Reptiles

Mark Catesby (English, 1682–1749)
THE BROWN CURLEW. Numenius fuscus. Tantalus fufcus. L. Arum sagittariae folio angusto, acumine & auriculis acutissimis. Arum fagittae folium. L.

POMEGRANATE BOX 6099 ROHNERT PARK CA 94927

Birds, Beasts & Reptiles

Mark Catesby (English, 1682–1749)
Bison Americanus. Bos Bison. B. L. Pseudoacacia hispida floribus roseis. Robinia hispida. L.

Pomegranate Box 6099 Rohnert Park CA 94927

ana Aquatica
Sarracena

BIRDS, BEASTS & REPTILES

Mark Catesby (English, 1682–1749)
THE WATER FROG. Rana aquatica. Rana esculenta. L.
Sarracena foliis brevioribus latioribus: Sarracena
Canadensis, foliis cavis & auritis. Sarracena purpurea. L.

POMEGRANATE BOX 6099 ROHNERT PARK CA 94927

Psitticus Caroliniense.

The Parrot of Carolina

Birds, Beasts & Reptiles

Mark Catesby (English, 1682–1749)
THE PARROT OF CAROLINA. Psittacus Caroliniensis. Psittacus Carolinensis. L. THE CYPRESS OF AMERICA. Cupressus Americana. Cupressus disticha. L.

Pomegranate Box 6099 Rohnert Park CA 94927

Anas cristata elegans.
The Summer Duck.

BIRDS, BEASTS & REPTILES

Mark Catesby (English, 1682–1749)
THE SUMMER DUCK. Anas Americanus cristatus elegans. Anas Sponsa. L.

POMEGRANATE BOX 6099 ROHNERT PARK CA 94927

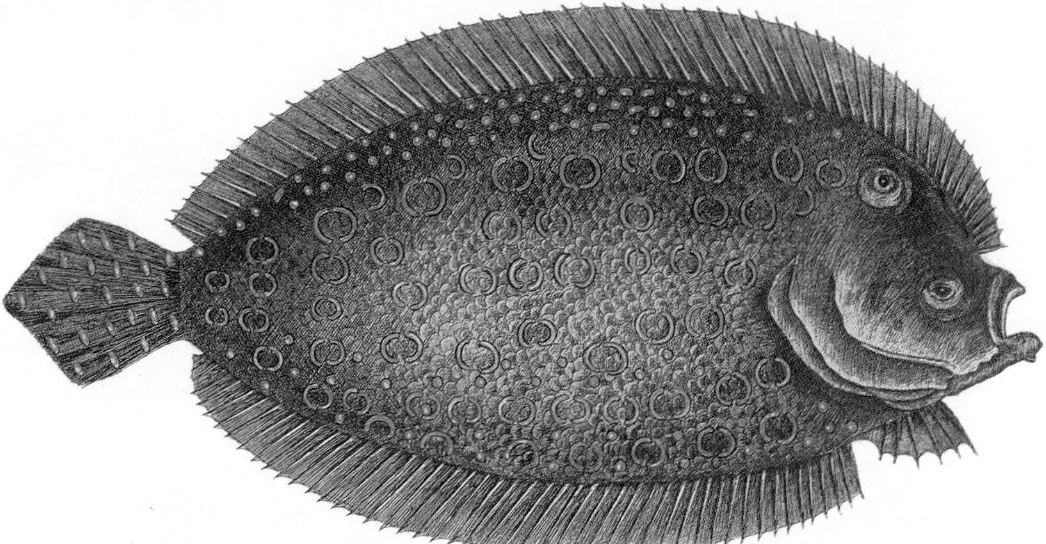

Solea lunata &c

Birds, Beasts & Reptiles

Mark Catesby (English, 1682–1749)
THE SOLE. Solea lunata & punctata. Pleuronectes lunatus. L.

Pomegranate Box 6099 Rohnert Park CA 94927

73
Phœnicopterus.
Keratophiton &c.

Birds, Beasts & Reptiles

Mark Catesby (English, 1682–1749)
THE FLAMINGO. Phoenicopterus Bahamensis.
Phoenicopterus ruber. L. Keratophyton dichotomum fuscum.

Pomegranate Box 6099 Rohnert Park CA 94927

Pica cristata cærulea
The crefted Jay.

BIRDS, BEASTS & REPTILES

Mark Catesby (English, 1682–1749)
THE BLUE JAY. Pica glandaria coerulea cristata. Corvus cristatus. L. THE BAY LEAVED SMILAX. Smilax laevis folio Lauri, baccis nigris. Smilax laurifolia. L.

POMEGRANATE BOX 6099 ROHNERT PARK CA 94927

Lilium &c. Anguis &c.

BIRDS, BEASTS & REPTILES

Mark Catesby (English, 1682–1749)
THE WAMPUM SNAKE. Anguis e coeruleo & albo varius. Coluber fasciatus. L. THE RED LILLY. Lilium Carolinianim, flore croceo punctato petalis longioribus & angustioribus. Lilium philadelphicum. L.

POMEGRANATE BOX 6099 ROHNERT PARK CA 94927

Birds, Beasts & Reptiles

Mark Catesby (English, 1682–1749)
UROGALLUS minor fuscus cervice, plumis alas imitantibus donata. Tetrao Cupido. L. Meadia. Dodecatheon Meadia. L.

Pomegranate Box 6099 Rohnert Park CA 94927

Œnanthe Americana. *Solanum &c fl: purpureo.*

Birds, Beasts & Reptiles

Mark Catesby (English, 1682–1749)
THE YELLOW-BRESTED CHAT. Oenanthe Americana pectore luteo. Solanum triphyllon flore hexapetalo, tribus petalis purpureis erectis ceteris viridibus reflexis. Trillium sessile. L.

Pomegranate Box 6099 Rohnert Park CA 94927

11

Birds, Beasts & Reptiles

Mark Catesby (English, 1682–1749)
THE TUMBLE TURDS. Scarabaeus pillularis Americanus. Scarabaeus carnifex. L. Scarabaeus. Scarabaeus pillularius. L. Lilium sive Martagon Canadense, floribus magis flavis non reflexis. Lilium Canadense. L.

Pomegranate Box 6099 Rohnert Park CA 94927

T. 84

Numenius &c.

Birds, Beasts & Reptiles

Mark Catesby (English, 1682–1749)
THE RED CURLEW. Numenius ruber. Tantalus ruber. L.

Pomegranate Box 6099 Rohnert Park CA 94927

T. 44
Avis Tricolor.
The Painted Finch.
itea Floridana.

Birds, Beasts & Reptiles

Mark Catesby (English, 1682–1749)
THE PAINTED FINCH. Fringilla tricolor. Emberiza Ciris. L. THE LOBLOLLY TREE. Alcea floridana quinque capsularis, Laurinis foliis, leviter crenatis, feminibus coniferarum instar alatis. Hypericum Lasianthus. L.

POMEGRANATE BOX 6099 ROHNERT PARK CA 94927

Birds, Beasts & Reptiles

Mark Catesby (English, 1682–1749)
Anona maxima, foliis oblongis angustis, fructu, maximo luteo conoide cortice glabro in areolas angulares distincto. Annona reticulata. L. Phalaena magna ex rufo & albo varia Americana. Phalaena cecropia. L .

Pomegranate Box 6099 Rohnert Park CA 94927

Icterus.
The Baltimore Bird.
Arbor Tulipifera.
The Tulip Tree.

Birds, Beasts & Reptiles

Mark Catesby (English, 1682–1749)
THE BALTIMORE BIRD. Icterus ex aureo nigroque varius. Oriolus. Baltimore. L. THE TULIP TREE. Arbor Tulipifera Virginiana. Liriodendron tulipisera. L.

Pomegranate Box 6099 Rohnert Park CA 94927

Birds, Beasts & Reptiles

Mark Catesby (English, 1682–1749)
THE PIED-BILLED DOBCHICK. Podiceps minor rostro vario.Colymbus podiceps. L.

Pomegranate Box 6099 Rohnert Park CA 94927

BIRDS, BEASTS & REPTILES

Mark Catesby (English, 1682–1749)
THE BLUEISH GREEN SNAKE. Anguis gracilis caeruleo viridis. Coluber mycterizans. L. Frutex baccifer verticillatus, foliis scabris latis, dentatis, & conjugatis baccis purpureis dense congestis. Callicarpa Americana. L.

POMEGRANATE BOX 6099 ROHNERT PARK CA 94927

T 86

An Thymelæa foliis obtusis. *Anseri Bassano &c.*

BIRDS, BEASTS & REPTILES

Mark Catesby (English, 1682–1749)
THE GREAT BOOBY. Anseri Bassano congener.
Pelecani Bassani pullus. L. An Thymelaea foliis obtusis(?).

POMEGRANATE BOX 6099 ROHNERT PARK CA 94927

uercus, an potius Ilex Marilandica
lio longo angusto, Salicis: Raii Hist.
Willow Oak.

Picus maximus rostro albo
Largest White Bill'd Woodpecker.

T. 16.

BIRDS, BEASTS & REPTILES

Mark Catesby (English, 1682–1749)
THE LARGEST WHITE BILLED WOODPECKER. Picus maximus rostro albo. Picus principalis. L. THE WILLOW-OAK. Quercus, An potius Ilex Marilandica folio longo angusto salicis. Quercus phellos. L.

POMEGRANATE BOX 6099 ROHNERT PARK CA 94927

Rana *Arum*

Birds, Beasts & Reptiles

Mark Catesby (English, 1682–1749)
THE GREEN TREE FROG. Rana viridis arborea. Rana arborea. L. Arum Americanum Betae folio. Dracontium foetidum. L.

Pomegranate Box 6099 Rohnert Park CA 94927

Cornus &c
An Phaseolus &c.
Orbis &c.

Birds, Beasts & Reptiles

Mark Catesby (English, 1682–1749)\
THE GLOBE FISH. Orbis lavis variegatus. Tetrodon lagocephalus. L. Cornus foliis salicis laureae acuminatis, floribus albis, fructu Sassafras. Laurus aestivalis. L. An Phaseolus minor lactescens, flore purpureo(?). An Clitoria galactia(?). L.

Pomegranate Box 6099 Rohnert Park CA 94927

Birds, Beasts & Reptiles

Mark Catesby (English, 1682–1749)
THE TURTLE OF CAROLINA. Turtur Carolinensis. Columba Carolinensis. L. THE MAY APPLE. Anapodophyllon Canadense Morini. Podophyllum peltatum. L.

Pomegranate Box 6099 Rohnert Park CA 94927

Tapia &c.

Cancer terrestris.

BIRDS, BEASTS & REPTILES

Mark Catesby (English, 1682–1749)
THE LAND-CRAB. Cancer terrestris, cuniculos sub terra agens. Cancer ruricola. L. Tapia trifolia fructu majore oblongo. An Crateva Tapia(?). L.

POMEGRANATE BOX 6099 ROHNERT PARK CA 94927

T. 39.

Coccothraustes caerulea.
The blew Grosbeak.

Magnolia Lauri folio, subtus albicante.
The Sweet Flowering Bay.

BIRDS, BEASTS & REPTILES

Mark Catesby (English, 1682–1749)
THE BLUE GROSBEAK. Coccothraustes coerulea. Loxia coerulea. L. THE SWEET FLOWERING BAY. Magnolia laurifolio subtus albicante. Magnolia glauca. L.

POMEGRANATE BOX 6099 ROHNERT PARK CA 94927

Suillus.
The Back Fin.

Birds, Beasts & Reptiles

Mark Catesby (English, 1682–1749)
THE GREAT HOG FISH. Suillus. Labrus maximus.
Forst. N.

POMEGRANATE BOX 6099 ROHNERT PARK CA 94927

Cornus mas &c.

Turdus minor &c.
The Mock-bird.

Birds, Beasts & Reptiles

Mark Catesby (English, 1682–1749)
THE MOCKBIRD. Turdus minor cinereo albus non maculatus.Turdus polyglottos. L. THE DOGWOOD TREE. Cornus mas Virginiana. Cornus florida. L.

Pomegranate Box 6099 Rohnert Park CA 94927

Vipera Fusca.
Arum &c.

Birds, Beasts & Reptiles

Mark Catesby (English, 1682–1749)
THE BROWN VIPER. Vipera fusca. Coluber luridus.
Forst. N. Am. Arum maximum AEgyptiacum quod vulgo
Colocasia. Arum Colocasia. L.

Pomegranate Box 6099 Rohnert Park CA 94927

T 39

BIRDS, BEASTS & REPTILES

Mark Catesby (English, 1682–1749)
THE HAWKS-BILL TURTLE. Testudo Caretta. Testudo caretta. L.

POMEGRANATE BOX 6099 ROHNERT PARK CA 94927

Sciurus *Viscum*

Birds, Beasts & Reptiles

Mark Catesby (English, 1682–1749)
THE FLYING SQUIRREL. In a flying attitude. Sc. volans. Viscum caryophylloides, Aloes foliis viridibus acuminatis, floribus racemosis luteis. An Tillandsia utriculata(?). L.

Pomegranate Box 6099 Rohnert Park CA 94927